Ann and the ISLAND STORM

By ELSPETH BRAGDON
Illustrated by MARJORIE TORREY

Originally titled *Fairing Weather*
First published in 1955
Cover illustration by Ecaterina Leascenco
Cover design by Robin Fight
© 2020 Jenny Phillips
goodandbeautiful.com

*Dedicated to keepers of lights of all kinds,
at all times, and in all places.*

The author wishes to thank Edward Rowe Snow for his generous permission to adopt and adapt material from his book, *Famous Lighthouses of New England*.

Contents

1. Ann's World 1
2. Here's Betsy! 7
3. Grandma Gates 16
4. The Big Storm 28
5. Who's a Scaredy-Cat? 39
6. School Is Fun! 53
7. Betsy Tries Hard 61
8. What's a Line Storm? 72
9. Come Kitty, Come Kitty! 80
10. Let's Be Friends 88

1

Ann's World

ONE DAY IN September, Ann Marsh sat on a wide, flat stone and hummed. The air was soft and warmish, with a surprising little nip of chill when the wind blew. It was Ann's favorite kind of day. Besides, it was her ninth birthday, and she had a special birthday feeling. It was a shining day, and there would not be many more, for soon the equinoctial storms would come, and the short gray days of winter would follow. So Ann looked at her shining world and hummed under her breath—not a tune, but more like a kitten's soft contented purr.

Ann's world, her very own world, was not very large. All around it, though, was the great, wonderful, changing sea with its stubbly little islands. In the distance the mountains of Big Island rolled up and over upon one another, like waves. Her very own world was Bleak Island, four miles from the shores of the state of Maine. It was a roundish island about a mile and a half wide, and, Ann knew, it had everything that can make an island wonderful.

The sea, of course, came first. Pale and still at twilight, it spread like liquid honey to distant shores. Deep blue and ruffled with whitecaps, it danced and chortled under a warm July wind. In the winter it beat with a thundery sound against the tall rocks to the east. It was never twice the same, for sun, wind, fog, or rain gave it different colors and sounds. Almost the nicest sound it made was in slapping against the seaweed at the end of the boat runway—not angry, but teasing and laughing—as it did today, on Ann's birthday.

Then there was the lighthouse of which Ann's father was the keeper. Most lighthouses are built on a rocky headland from which the light can shine out far over the dark waters. But the Bleak Island light tower was built right in the water, its base on one of a series of long granite reefs that stretched out into the sea. You reached it by a sort of bridge, or gangway, across the swirling water, and when you had crossed over, you could climb the winding stairs and go out on the little catwalk that circled the tower. You were then so high that you could see for miles. The seagulls wheeled over your head as they might if you were some strange bird.

There were two houses on Bleak Island. One was the white-walled brick house where Ann lived. It stood on a little rise of ground with a wall of blue-black pines in back of it. A little way to the east, there was a grassy place where the keeper's cow grazed. Still farther to the east was Mrs. Gates' house.

Mrs. Gates was a very old lady. She lived all alone, for her grown-up son was married and lived in Ohio. Her husband had been keeper of the light for many years until he died. At that time Mrs. Gates really should have left Bleak Island, which was government property. But she had refused to go and had sent over to Big Island for enough sawed lumber to build the little yellow house where she now lived. Ann's father used to laugh and say, "It would take more than the President of the United States and the whole United States Navy to move Perley Gates' widow."

Ann thought the President of the United States was too smart to try. She was a little afraid of Mrs. Gates, who could speak sharply if she'd a mind to, but the old lady was actually more lonesome than cross. There was one very fine thing about her.

She had a big cat named Lady Topside, and Lady Topside was a great one for having kittens. One of them had been given to Ann for a birthday present.

On the north shore, there was a little pool among the pink rocks. When the tide came in and the sun warmed the water in the rock pool, it was a fine place for wading. Farther to the west, on the north shore, was Picnic Point, to which the summer people came. They lugged heavy baskets of food across the meadow and built smoky fires on top of the cliff. They were there today, and Ann could hear them laughing and singing. She could smell the boiled lobsters. But it never, for one moment, entered her head to go near Picnic Point when the summer people were there.

Summer people, Ann had decided long ago, weren't people at all! She didn't know why she thought this was true, but she had gathered it from the coolness in her mother's voice when they stopped to look in at her kitchen door. There was an extra politeness in the way her father answered their questions when they visited the light. They just didn't know anything at all, Ann concluded. Picnic Point could be theirs for a few weeks, but they had better stay where they belonged and not start discovering her secrets.

A puffy white cloud slid across the sun, and Ann

suddenly felt cold. She stood up and turned to go into the house, but she was stopped by a steady thumping sound, coming nearer and nearer. Across the water from the west came a big gray boat that Ann knew well, and the thumping sound came from its powerful diesel engine. It was the Coast Guard boat from the base on Big Island. These boats brought supplies and coal and kerosene to Bleak Island, and their coming was always fun, for the crew was very friendly.

Ann flew down the path until she reached the top of the boat runway. The big boat dropped its anchor near shore, and two men got into the tender. Then a third man handed something down very carefully. Was it something or somebody? Ann's eyes grew very big.

It was somebody. It was a little girl, just Ann's size. She was all dressed up in Sunday clothes. As the rowboat came in toward shore, Ann could see a small pale face, with big eyes staring straight ahead. She saw that the little girl had a pocketbook in her hands, and on her hands were gloves.

"Hi, Ann," called one of the men, waving at her. "See what we've brought you!" But Ann didn't answer. She spun around on her heel and ran up the path to the house. She ran through the kitchen and upstairs like a frightened animal.

Ann's mother came out of the pantry, wiping her hands. "Ann Marsh," she cried sharply, "what's got into you?"

There was no answer but the sound of Ann's door slamming shut.

2
Here's Betsy!

MRS. MARSH KNEW that whatever had stirred Ann up was out-of-doors. She went to the window and saw two men from the Coast Guard boat coming up the path. They wouldn't upset Ann. She knew Bert Stanley and Iva Gilley as well as if they were her own brothers.

Then Mrs. Marsh saw that they were carrying a lot of packages and two suitcases. And after them, stumbling on the rocky path, came a little girl in city clothes—a little girl with big unhappy eyes in a pale face. "Goodness," Mrs. Marsh muttered to herself. "A child! Whatever in the world?" Then she saw that the men were heading for the kitchen door, and she ran to open it.

The men came in and put down the bags and packages. Bert put his big red hand on the little girl's shoulder. "This here little girl is going to Mrs. Gates' house," he said. "She's come a long way, and she never saw a boat before, or an island, either. I don't think she liked the trip over from Big Island very much."

The little girl stood very still in the warm kitchen and stared at Mrs. Marsh without smiling.

My, thought Ann's mother, *she's going to cry in about one minute.* She put her apron on briskly and said, "Come on, Bert and Iva, sit down at the table, and have a cup of coffee and some doughnuts. I'll just make some cocoa for—" Then she broke off and exclaimed, "Heavens to Betsy, I don't even know

your name, do I?" Then she gently began to take off the little girl's coat and hat, looking into the child's face lovingly.

"It's Betsy," said the little girl in a small voice.

"That's a good one," Bert laughed. "You said 'Heavens to Betsy,' and her name's Betsy!" And he clapped his hand on his knee.

Betsy put her hand into Mrs. Marsh's as if his loud laughter frightened her. Ann's mother picked Betsy up and held her close for a minute. "You're light as a feather, child," she said. "Now I'll put you down in your chair at the table, and you can hold the kitten while I make your cocoa. You'd like that, wouldn't you?"

A very small beginning-of-a-smile came to Betsy's face. When the kitten settled down on her lap and began to purr, the little girl gave a deep sigh, as if she weren't afraid anymore. Then she began to pet the small ball of fur very softly and gently.

When Mrs. Marsh went into the pantry to get the cocoa, Bert stepped lightly after her. Then he said in a whisper, "We brought her here first because we don't believe her grandma rightly expects her. No one told us she was coming. The bus driver brought her to the Coast Guard base, and he got her from the Boston train, and I guess someone in Boston must have got her from the Ohio train. For that's where she's from."

"Why, she must be Lou's child," said Mrs. Marsh.

"That's right," said Bert. "I don't think she's ever seen her grandma before. What she's doing, traveling all over by herself, I don't know. But I kind of hated to walk in on Mrs. Gates with her, not knowing how the old lady would take it."

Mrs. Marsh asked no questions but quickly started the cocoa cooking and poured out two cups of coffee from the pot on the back of the stove. Then she went to the foot of the stairs and called Ann, but there was no answer.

When the cocoa was done, she poured out a big cupful for Betsy and put a marshmallow in it. Then she put a plateful of doughnuts and hermit cookies on the table. Getting into her coat, she said to Betsy, "Dear, you take care of the kitty for a few minutes, and drink that nice cocoa. I've got an errand to do, but I'll be right back. Bert and Iva, you just make yourselves at home. Turn on the radio if you want to."

Betsy looked up. "Are you going to the store?" she asked.

Ann's mother laughed. "There's no store on this island to go to," she said. Then she left, without seeing the faint shadow of doubt on Betsy's face. What kind of place could this be, the child wondered, where there was no store?

Mrs. Marsh ran out of the kitchen in the fast-falling twilight, hurrying along the path through the tall cedar trees, and in no time at all was knocking at Mrs. Gates' door. The old lady welcomed her warmly, and they were soon sitting together in the kitchen.

"Mrs. Gates," began Ann's mother, "I have rather a surprise for you." Then she told her all that she knew about Betsy's arrival, ending, "And right at this minute that little thing is sitting in my kitchen, drinking cocoa and playing with the kitty, trying not to show how strange and lonely she feels!"

Mrs. Gates got up slowly and began to put on her coat and shawl. Her hands were shaking from excitement. "That's Lou's child, all right," she said. "Something must have happened to him. The last letter I had from him came in July, and he didn't say a word about sending the child to me. I'm awful glad to have her, but why in the world would he send a little mite of a thing clear across the country all by herself like that?" She picked up a flashlight and went to the door. "I'm going to fetch my granddaughter," she said in a trembling voice.

Ann's mother put her hand on the old lady's arm. "She's tired, and she's scared, Mrs. Gates," she said gently. "Let's not ask too many questions tonight. Let's wait a little. We don't want her to start crying and carrying on."

Mrs. Gates nodded and tied a shawl over her head. "I'll wait," she promised. Then they started down the path together.

An hour later, Mrs. Marsh started up the stairs to Ann's room. Darkness had come. The big light in the tower was blinking steadily, and the kitchen was filled with the smell of good things cooking for supper. Betsy had gone with her grandma, holding fast to her hand. Bert and Iva had gone off on the Coast Guard boat. In a few minutes, Captain Marsh would come in from the lighthouse. Now she must find out what was wrong with Ann. There must be something wrong, for there hadn't been a sign or sound from her during all the excitement.

The room was dark when Mrs. Marsh opened the door. When she switched on the light, she thought for a moment that the little girl was asleep, for she lay facedown on the bed, very still. She sat down on the edge of the bed and touched Ann's shoulder gently. The child looked up, her face hot and flushed, as if she had been crying.

"I heard," said Ann in a sort of choking voice. "I heard everything Bert said. I saw her, too. I don't want her to come here, not one single bit. She's one of the summer people. She doesn't belong here. She wears funny clothes. She won't know anything. She's a scaredy-cat, too. I wish she'd go away."

Mrs. Marsh looked at Ann in a shocked way, as if she hardly knew her. Then she said slowly, "Ann, I'm really ashamed of you. What's all this talk about summer people? They're just people, aren't they? Don't ever let me hear you talk like that again." Then she smoothed Ann's hair back from her hot face.

"Besides," she went on, "Betsy isn't 'summer people,' whatever you think that means. She comes from far away, of course, but so did I once upon a time. How do you think I'd have liked it, when I came to Big Island from Massachusetts to teach school, if folks hadn't been friendly to me? Betsy's come to visit her grandma. She's far away from her father and mother. She needs folks to be friends. Now I want you to wash your face, and come downstairs to supper."

She stood up and went to the door. Then she turned. "Come to think of it," she said, "if your dad hadn't wanted to be friends with me when I came here, you wouldn't be having a birthday cake tonight, would you?" Then she smiled such a nice twinkly smile that Ann had to smile back.

The little girl got up slowly and went toward the

bathroom. With her hand on the doorknob, she stopped. "Have I got to be best friends with her?" she asked in a low voice.

Her mother smiled. "Nobody is best friends with anyone right away," she said. "You just try being plain friends first. She's a real nice little girl, I think. The kitty sure liked her. Got right into her lap and purred."

"Hmm," said Ann, scrubbing her face. "Well, I'll be plain friends. She can't be too bad if kitty liked her." Then she ran downstairs as fast as she could to see her birthday cake.

"Happy birthday!" shouted her father. He picked her up and kissed her ten times, once for each birthday and one to grow on!

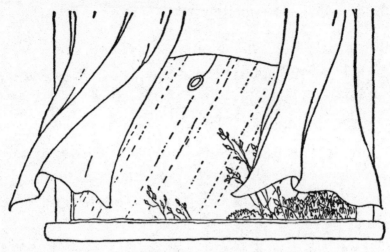

3
Grandma Gates

EARLY THE NEXT morning, Betsy woke suddenly and sat up in bed, staring around her. For a long moment, she couldn't imagine where she was, nor could she remember how she had come to this small room over which the roof slanted steeply. This was not her bed, nor her room in Ohio. None of the sounds she heard were familiar. Instead of the honking of automobile horns and the clatter of trucks, she heard the steady splashing of rain on the roof so close to her head. She heard, instead of the voices of other families in the apartment house where she had lived, a low, steady roar, like far-off thunder. The air in the room was cold, and there didn't seem to be any radiator to turn on.

Quick as a flash, she jumped out of bed and ran to the window. Then she shut it hard to keep out the rain and wind. She put on her warm bathrobe and slippers. She knew now where she was. She was at her grandma's house, far from her father and mother and everything she knew best. She looked down the steep stairs to a patch of warm light at the foot. "Grandma, where are you?" cried Betsy in a quavery voice.

Mrs. Gates, a white apron tied around her waist and a big spoon in her hand, hurried to the foot of the stairs. She was smiling widely as she waved the spoon at Betsy. "Now, you're what I call a real fine girl," she said. "You're up early, rain or no rain. I thought maybe you'd want to sleep late, this once, after your long trip. I'm frying doughnuts. You come down right this minute, just as you are, and eat the holes for me." Then she hurried back to the kitchen.

Betsy came downstairs slowly. What did Grandma mean by "eat the holes"? You can't eat holes!

Two big oil lamps with bright tin reflectors were lit in the kitchen, a kettle hummed busily on the stove, and Lady Topside was busily licking the smallest of her new batch of kittens. Betsy stopped shivering and sat down on a small stool near the old mother cat, who paid no attention to her at all. All of a sudden, Betsy heard a loud, sharp cry. "Grandma!" she asked excitedly, "what's that sound? Is someone hurt and crying?"

Mrs. Gates turned from the stove with a small plate in her hand. She gave it to Betsy before she answered. "Goodness, child," she said. "Don't you know a seagull when you hear it? That's a seagull, that is. He isn't hurt. He's calling to the other gulls, saying 'Fish! Fish! Fish!' He can see through the rain and fog, and even the water, wherever fish are, and he just screams his head off all the time. Now you eat up these holes, and I'll fetch you a glass of milk and an apple."

Betsy looked at the plate. On it was a pile of crispy golden-brown little balls, each just the size of the hole in a doughnut. She touched one carefully. It was hot as hot. Then she licked a crumb from her finger. It was simply delicious. Then she put a "hole" into her mouth, and it was as light and crunchy and good as cake, only perhaps better. "M-m-m!" she said. Then she ate every single hole and every bit of the apple and drank up every bit of the milk her grandma had brought her.

Grandma Gates pushed the kettle of fat to the back of the stove. She came over and sat down beside Betsy. She picked up one of Lady Topside's kittens and put it in the little girl's lap. Then she said gently, "It's real nice to have company, Betsy. I've been alone for a long time, ever since your pa went off to college. How is your pa, by the way?"

Betsy didn't answer for a moment. She felt the tears coming. She winked them back hard. "My daddy's sick," she said slowly. "If I knew where my coat was, I'd give you the letter Mommy gave me for you. I forgot it last night."

"It's behind the kitchen door," said Grandma Gates, rocking steadily. "You give me the kitten and fetch me the letter, and we'll read it together."

Betsy got the letter and brought it back to her grandma, who opened her arms. "There's lots of room in my lap for you," she said. "And there's room in your lap for kitty. We'll read the letter together, all three of us."

Betsy hated that letter which her mother had pinned into her coat pocket for safekeeping. She hated it and was scared of it. But somehow, sitting in Grandma Gates' soft, warm lap made everything seem not so bad.

Grandma opened the letter and read aloud slowly:

"Dear Ma Gates, I wish Lou and I had written you more often and that there had been time to let you know how things are before Betsy gets to the island. But there has been so much to do that

I'm about crazy, and time has just flown. To begin at the beginning—Lou came down sick last week, and the doctor said he had to go to a hospital out in Arizona for a while, maybe for a year. The folks he worked for were real sorry about it and have been good to us. They gave me a job in the office to help out on the hospital bills. We had to give up the apartment and put our furniture in storage. I got a furnished room, but it's no place for Betsy to be, with no one at home when school is over. So we thought maybe you'd take her for a while. We'd never forget it if you would. At first I thought I couldn't stand it when Lou got sick, but the doctor says he has a good chance of getting well, and that's all that matters. Betsy is a real good girl and won't be any trouble once she gets used to living out of a city. So thank you a lot, Ma, and I'll write more often when things get straightened out."

Betsy had cried a little as Grandma read, for the letter sounded just like her mother talking. She remembered the day she went to the station with her mother to say goodbye to her father. He had looked real sick. He had smiled as if it hurt to smile. It had been just awful!

Grandma Gates put her hand under Betsy's chin and tipped back her face until she could look into the little girl's eyes. Then she gave her a great hug. "Betsy," she said, "you and I are going to have a real fine time. You wait and see. When you get tired of playing with me, you'll have Ann to play with. You'll be friends with her ma and pa, and Bert and Iva and Stan, and all the other folks who come here. You just remember your pa is doing his best to get well and your ma is working hard to help him. You and I will help both of them by having a real fine time together."

Then she gently eased the little girl out of her lap. She stood up and, taking Betsy by the hand, walked over to the sink by the east window. "Let's do the dishes now," she said cheerfully. "You wipe and I'll wash. Then I'll bring your clothes downstairs and you can dress right by the fire. Then we'll make plans."

Betsy smiled uncertainly. It sounded like fun the way Grandma said it. She reached for a dish towel.

When the dishes were done and Betsy had dressed by the fire, Grandma said briefly, "Beds!" Betsy trudged upstairs to her room. It seemed very cold compared to the warm kitchen. She hung up her pajamas and bathrobe, yanked up the bedcovers, and straightened the pillow. Then she went to the window and looked out into a rainy, wind-swept world. The long grass between Grandma's house and the shore was blown flat under the rain, and the sea, which was a cold, miserable-looking gray, rolled by in big rough waves edged with white. She could hardly see the long line of hills on Big Island, and no boats were to be seen anywhere. It was a horrible, lonesome day, thought Betsy.

Just then she heard Grandma's step on the stairs. A moment later she appeared at the door. She looked at Betsy's bed (which did look somewhat lumpy), marched over, and pulled off all the bedclothes. "Betsy Gates," she said briskly. "You don't call that making a bed, do you? Come on over here, and we'll do it the right way."

Betsy didn't move. "That's the way I like my bed to be," she said in a low voice.

Grandma tugged at the mattress. "Come here, child," she said again. "Help me turn this mattress over. There's no royal road to bed-making. You either do it right, or the bed isn't made at all."

Together they turned over the mattress, put each sheet on smoothly, and made neat tight folds at the corners. Then the pillow was shaken and punched and smoothed and put at the head of the bed. A many-colored cover was spread over all, and there it was, neat as a pin!

Betsy looked at the bed. Then she looked at her grandma. "Do I have to do all this every single day?" she asked in a cross voice.

"Indeed you do," Grandma answered, laughing at the little girl's expression. "There's no royal road—"

"What's a royal road?" interrupted Betsy.

Grandma smiled. "That's a saying my ma had," she replied. "A royal road means a nice, easy road for a lazy king to ride on. When you say there's no royal road to anything, you mean there's no easy way to do it. Now, you unpack your clothes, and put them away. Then come down to the kitchen."

As she turned toward the door, Betsy put her hand on the bright coverlet on her bed. "What do you call this, Grandma?" she asked. "Is it a spread?"

Grandma shook her head. "No," she said. "That's a patchwork quilt. I made it when I was seventeen, the winter before I married your grandpa. Maybe you'll make one someday. They're warm and pretty, and they last a long time. The quilt on my bed is 'most a hundred years old. Older than me," she added, laughing. Then she hurried downstairs.

When Betsy had unpacked her suitcases and boxes, she went downstairs, carrying her doll, Diane. The smallest kitten was curled up in the rocking chair, so she settled Diane beside it and looked around the kitchen. Somehow it didn't seem as bright and cheerful as it had seemed earlier.

"Grandma," she said, "there's nothing to do!"

Grandma came out of the pantry carrying a queer

flat thing. It looked a little like a fish, but it was as stiff and dry as a board. "I'm going to pick out this codfish and set it to soak," she said. "You set on that little stool by the stove, and look over the pieces of cloth in my sewing basket. Maybe you'll find enough to make a quilt for that doll baby of yours. Want a piece of codfish?"

Betsy looked at the hard, boardlike fish and shook her head.

"It's good," said Grandma. She took a knife and sliced off a little and put it into her mouth.

Betsy put out her hand. "I'll have some," she said uncertainly.

Betsy sat on the stool and looked over the pieces of bright cotton. She noticed that Grandma hadn't swallowed her piece of codfish but was chewing it the way you chew gum. The little girl copied her, but she couldn't make up her mind whether she liked this strange leathery stuff or not. At first it tasted a little the way glue smells. It was awfully salty. But it didn't seem polite to spit it out, so Betsy chewed away cautiously, and after a while it tasted better.

Just then someone knocked at the door.

Mrs. Gates called, "Come in," and the door opened with a great whoosh of rain and wind. There was Ann, dripping wet, her face very rosy from the cold.

"Hi," said Ann. She pulled off her rubber boots and coat and put them behind the stove to dry.

Betsy watched her silently, feeling rather shy. Ann was just her size, but she hadn't minded the storm outside. She must be very brave and strong.

Grandma Gates said briefly, "Betsy, this is Ann Marsh. You two can have some high old times together."

The two little girls looked at each other unsmilingly.

Ann really wasn't so sure about this newcomer. Betsy looked pale and quiet and almost scared. Ann didn't think much of scaredy-cats. However, she remembered what her mother had said about Betsy's being far away from home, so she crossed the room. She looked at the black kitten. She looked at the doll. Then she looked at the pieces of cotton Betsy was holding.

"What's that cloth for?" she asked gruffly.

"I'm going to make a quilt maybe," said Betsy timidly. "Grandma's going to show me how to make a little quilt for my doll. I'm looking over pieces."

Ann sat down on the floor and began to rummage in the sewing basket. *Oh dear,* she thought, *Betsy is nothing like me.*

Grandma Gates looked over her shoulder at both of them. "The quilt on Betsy's bed I made when I was seventeen. I began looking over pieces for it on a day just like this—a mean, wet, tedious day. I remember thinking I'd never in the whole world finish a big quilt, and I didn't rightly want to. I was a wild one, like Ann. I wanted to be like the boys and keep a light. I didn't want to sew and keep house. I didn't even want to sit still. Well, I didn't sit still long, I can tell you."

The little girls said in one voice, "What happened?" Then they looked at each other and smiled.

Grandma Gates put the last of the codfish to soak before she answered, and then she picked up the kitten and the doll and put them on the floor beside Betsy. She took her knitting from a bag and settled herself comfortably. She began to rock as she knitted. Then she said slowly, "Well, it was like this . . ."

4
The Big Storm

I WAS NINE YEARS old when my pa was sent to Misery Light. Whoever named that point named it well, for it was so rocky and mean that not even bushes or trees would grow there. It was just a sort of ledge sticking out from the land. When the tide came in, it was an island, for the water came right up over the narrow road that led to town eight miles away. I hated it when I first saw it, but I came to care an awful lot for that hump of rock before I left it to marry your grandpa.

"There was just Pa and my two brothers, Ed and Nathan, besides me; my ma had died the winter before. I had to keep house for the four of us, and it was a job too. I cooked and washed the clothes; I mended and scrubbed the floors. I fed and cared for the hens, which were my pets. After a while I came to know every bit of that ledge by heart, and, whenever I had any free time, I ran all over it like a wild animal, climbing the steep rocks and exploring the caves under them. In the evenings I'd read the old log books which lightkeepers before my pa had written. When I read about some of the real bad storms they'd had, I was glad we lived in a good strong house built of rock like the ledge.

"Pa and the boys were busy night and day, tending the light, keeping things shipshape, and, in bad weather, watching out for vessels in trouble.

"Our lighthouse tower had thirty lamps in it, all oil lamps, and each had to be lit by hand. The chimneys and reflectors had to be kept shining clean. Just a short way from the light, there was a bell tower, and the bell in it weighed twelve hundred pounds. The bell was rung by clockwork, and, during a foggy spell, the clockwork had to be wound every five hours. We all worked together, and we all worked hard, because we knew how treacherous the coast was to the east and the south of us, and how mean the tide was.

"One day not long after Christmas, the year I was seventeen, Pa decided to go to town and lay in supplies. As the tide was running high, he set out in his dory, for the sea was still and glassy. We watched him set off, and, as he waved goodbye, he shouted, 'I'm counting on you to take care of things.'

"Ed and Nathan weren't a lot older than me, but they were big and strong. I had no reason to feel scared about being left with them, but there was something about the pale gray sky and the oily-looking water that made me uneasy. The boys went to the light, and I went back to my chores. As I went, I wondered what we would do if a bad storm came.

"I didn't have long to wonder. By noon the sky was dark, and by four o'clock the snow was falling quickly. The boys had lit the lamps and were watching the angry waves from the tower. The wind blew harder and harder, so I fastened the heavy wooden shutters tightly over all the windows.

"At six o'clock Ed came running through the darkness. 'We can't come up for supper,' he shouted above the howling wind. 'Fix me up something hot, and I'll take it back to the tower. Then you bar the door.'

"I gave him a kettleful of soup and a box of hard crackers, and he rushed out again into the night. I was just about to bar the door, as he told me to do, when I thought of my hens. I lit a lantern and tied a shawl over my head and put on Pa's oilskin coat and boots."

Grandma put down her knitting and looked out the window in a remembering sort of way. "The wind tore my shawl off the minute I stepped outside," she went on. "By the time I reached the chicken coop, the high rubber boots had filled up with snow, so I could hardly lift my feet. The oilskin coat had been ripped from stem to stern by the wind. I grabbed four hens by the legs and ran back to the house. When I got there, I'd have given a lot to be able to stay right there, but four more hens were in that coop. Back I went and got them and, somehow or other, got back to the house. Then I pushed the door to and barred it before I took off my wet clothes."

Grandma paused and took up her knitting before she continued. "That was a real bad night," she said. "The waves hit our old stone house so hard that it shook as if it were made of cardboard. Every now and then a window would crack upstairs where there were no shutters. Once, the chimney to the stove in the front room toppled under the force of the wind, and the bricks clattered on the roof like thunder. Seawater came in under the door sills until I stopped the cracks with rolled-up rag rugs. It was about the longest night I ever knew. I didn't dare let myself worry about Pa or the boys or people in boats in that storm. I just stitched away at my quilt

and kept the fire in the stove going, grateful that the woodbox by the stove was well filled.

"Daylight was a long time coming, and the sky had just begun to lighten a little when there was a loud knock at the barred door. I ran to it, and there were Ed and Nathan, each of them half-carrying, half-dragging a strange man, whom they let down gently on the floor near the stove.

"'Don't ask questions now,' said Ed. 'Give us some good hot coffee, and fill the big tin tub with hot water. Then heat some bricks, and light the stove in the spare room. Get the bed ready there, and put plenty of blankets on it.'

"I hustled then, as you can guess," Grandma said. "While I was fixing the spare room, the boys stripped off the wet clothes from the strangers and gave them hot baths. They were dressed in old but warm clothes of Pa's and were drinking coffee by the time I came back for the hot bricks to slip between the bedsheets to warm them. After the men had been put to bed, the two boys sat down, at last, by the fire and pulled off their heavy wet sweaters and socks. As they drank their coffee, I saw how tired they looked. But I couldn't wait any longer to ask what had happened.

"'It was this way,' said Ed slowly. 'About an hour ago, the light picked up a small vessel that was in trouble about fifty yards from shore. There were two men on her, and they were hanging onto the boat, which kept yawing over each time the waves hit her. A real big wave came along, and over she went. When the light hit her again, we could see the men were still there, holding onto her bow. Nathan and I ran down from the tower. He was giving the orders because he's older, so when he tied a rope around his waist and told me to hold onto the other end, I did so, though neither of us is much good at swimming. Then he waded into the waves and started to swim toward the boat.'

"'There was an awful bad surf,' Nathan took up in a tired voice. 'Half the time I didn't think I'd make it. It took me most of half an hour to reach them. When I got alongside the vessel so I could hang onto her, I could see one of the men was barely conscious. I yelled to Ed to pull on the line, and I grabbed the man as best as I could to hold his head out of the water. Then I told the other man to swim along with me to shore. He was in pretty bad shape too, but the waves helped bring us all in and piled us all up on the rocks. The first man landed right athwart me, and I couldn't have got free if Ed hadn't hauled him off. Then we had to roll them over and pump the water out of them and get enough life into them so we could help them up to the house.' As he finished, he put his head down on the table.

"I gave him a little shake. 'You can't go to sleep now,' I said. 'You've got to get some dry clothes on yourself and eat something. The wind's dropping some, I think. Then you can rest maybe.'

"By the time the boys had eaten and were warm and dry, day had really come. The wind had let up, though the surf still dashed high against the shore. The boys stretched out on the kitchen floor and slept as if they were dead. They couldn't sleep long, I knew. I put on my heavy coat and slipped out of doors to watch for them while they slept.

"What a sight I saw! The wooden passageway between the house and the tower was gone. So were the chicken coop and the small building we used for a storehouse. The little coal house stood on one end, held down by the fifty tons of coal stored there. Shingles were blown off the east side of our roof.

"I went up into the tower. The storm glass was gone, and everything was soaked with water and spray. I put out the lights and took out the lamps to clean them. The sea had dashed so high that there were pieces of driftwood and seaweed even on the catwalk around the light.

"As I looked down at the rocks below, I could see the broken planks of the little fishing boat that had been dashed to pieces not long after Nathan got the men from it safely ashore. Then, very faintly,

I heard the sound of a powerboat. It came nearer and nearer. It was headed for Misery Light. I never heard a more welcome sound in my life."

Grandma got up and looked out the window. "My pa was in that boat, and with him was the young man who was your grandpa later on. I ran to meet them, and, as soon as they had landed in Pa's dory, which they had brought with them, I told them how brave Nathan and Ed had been.

"Pa looked around at the broken windows, the smashed fishing boat, and the scraps of shingles from the roof. 'They're good boys,' he said slowly. Then he looked at me and smiled. 'What did you do in the storm?' he asked.

"I felt shy and almost ashamed. I couldn't think of anything brave I had done. Then I remembered. 'I saved the chickens!' I said proudly.

"My pa laughed loudly and then gave me a great hug. 'You're my brave girl,' he said. 'Now let's go up to the house. If those strangers feel all right after a while, Ben here can take them ashore.' And so he did."

Ann and Betsy had stopped sorting pieces early in the story and just sat staring at Grandma, trying to imagine her as a young girl long ago.

"Well," said Grandma. "The rain has stopped, and the wind has dropped quite a bit. There's a little piece of blue sky to the north, and I believe it's going to clear. Why don't you two go outdoors now and get some fresh air?"

Betsy hung back. It still looked cold and wet outside. The kitchen was warm and cozy. But Ann ran to get her coat and pull on her boots. "Come on, Betsy," she said cheerfully.

Grandma came from the hall closet, where she had been rummaging. In her hands she held a small oilskin coat and some boots. "This shows what comes of not throwing things away. These were your pa's when he was no bigger than you."

Betsy stood still, saying nothing.

Ann jumped up and down impatiently. "Come on," she urged. "We'll play lighthouse keeper in the big storm." Then she paused. "You can be Nathan," she said.

Betsy put on the boots and coat. Then she said in a small voice, "I'd rather be a hen!"

Ann laughed and laughed. So did Grandma Gates. "All right," said Ann. "I'll be Nathan."

Out they went, shutting the door behind them.

"Hmm," said Grandma Gates at the north window. "The sun's coming out real soon."

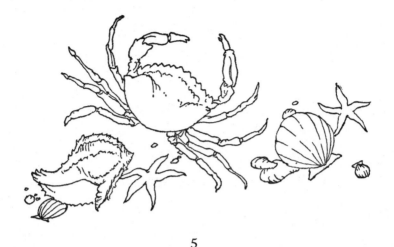

5
Who's a Scaredy-Cat?

THE AFTERNOON OF the first day Betsy spent on Bleak Island was about the most mixed-up time she had ever known. It could have been lots of fun, but it was a queer place, she thought. It was hard to walk in the heavy rubber boots, for one thing. But the tall sharp grass was still heavy with rain, and the blueberry bushes were hung with bright drops of water, so she could see why rubber boots were a good thing. She trudged along over an overgrown path as Ann ran ahead of her, pushing back branches of cedar trees that hung down wetly. In a very few minutes, Ann was out of sight, so Betsy was alone when she came to the first strange thing she was to see that day.

Just off the path, there was a low fence over which a few vines grew. The fence went around a small patch of bright green grass, very different from the coarse copper-colored grass that edged the path. On that square of green stood three stones. Two of them looked very old, and they leaned over sideways as if they were tired. The third was a rounded stone, the color of the pinkish rocks that were along the shore.

Betsy carefully climbed up on the fence near the leaning stones. She slowly spelled out the words that were carved on them. "JOHN STEBBINS," she read. "BORN MAY 6, 1797. DIED SEPTEMBER 12, 1839." Then, below, there were smaller letters, half hidden by the grass. "*May guardian angels watch over the sleeper*," Betsy read. The stone beside it had carvings, too. They only said, "MARTHA, WIFE OF JOHN STEBBINS," and the dates when she was born and when she died.

Betsy looked all around. The clouds had broken to let the sun pour down, and the sea beyond the Point was blue and sparkling. Three very white seagulls circled overhead against the deep blue of the sky.

"I think this is a good place for angels," she said to herself. Then she edged along the fence until she could see the round pinkish stone. All the carving on that stone said was one word: "GATES."

"Goodness gracious," said Betsy aloud. "That's my name."

"That's your grandpa's stone," said Ann, coming back to her. "He was Captain Gates, and he kept the light here for an awful long time. Your grandma put that rosebush there, and it looks real cheerful." Sure enough, just in front of the pinkish stone there stood a small rosebush on which a late rose nodded and danced in the wind.

"Who was John Stebbins?" Betsy asked.

"He was the very first man who came to this island to live," Ann said. "There wasn't any lighthouse here then, and there weren't any people for miles along the coast of Big Island, either. He and his wife were in a boat that was wrecked, and he rowed her in a small boat until he came ashore here. My daddy says he doesn't know how he ever got ashore here in any kind of storm. Come and see the big rocks that are just ahead. They'd smash a boat all to pieces if it came in close during a storm!"

The two little girls scrambled up the little rocky hill that led to the part of the island farthest to the east. All of a sudden, the path ended at a big flat rock with no grass or bushes on it. Ann took Betsy's hand. "I'll hold on to you," she said importantly. "We'll go look over the edge."

When Betsy got to the edge of the rock, she was very glad indeed that she had hold of Ann's hand, for down, down, down went the rocks to a great green-and-white swirl of water that rolled over and over in a great hurry to hit the rocks and break into spray.

"Oh my!" cried Betsy, pulling Ann back from the edge.

"Don't be so scared!" cried Ann. "Look at the seaweed!"

Betsy looked over and saw a lot of slippery brown stuff that swirled in the water as it rushed up and sank back. No matter how often the waves came, the shiny stuff clung to the rocks on which it grew, just letting its long streamers slide back and forth in the waves. Then she saw something round and pink. "What's that?" she asked, pointing.

Ann felt very impatient with Betsy. "Haven't you ever seen one of those before?" she cried. "That's a crab. There are hundreds of them along the shore."

Betsy blinked back the tears. She suddenly wished she were at home in Ohio where there weren't so many things she didn't know about. Just then, she heard a low snorting sound. It came from around the Point, just out of sight. Ann turned and ran toward it, Betsy following. In just a few minutes they came to another high place looking down on a long ledge of rocks. On it was a crowd of strange creatures stretched out on the seaweed, snorting and puffing lazily, as if to say, "Oh, how nice the sunshine is after all that horrid rain!"

Betsy stood and stared. Then she turned to Ann. "You didn't like it when I asked about the crab," she said slowly. "You said I didn't know anything. But I know what those animals are. I saw some just like them in a circus. They're seals."

Ann looked surprised. "I thought seals had to live in the ocean," she said. She was silent for a minute. Then she asked in a low voice, "What's a circus?"

Betsy smiled. "Let's take off these oilskin coats and sit on them, and I'll tell you about a circus. Did you know a seal could balance a ball on his nose?"

For a long while, the two little girls sat on the high point while Betsy told about the big circus tent and the trained lions and the pretty girls who flew high overhead on trapezes or walked a wire as easily as if it were a broad street. She told about the band and the clowns and the elephants. Ann forgot that she didn't like to answer questions herself and asked a good many, until she had a pretty good idea what a circus was like.

Then Betsy looked at the seals. "The seals on the rock are lots of different colors," she said. "Those in the circus were shiny black." She really didn't ask a question but waited to see if Ann could explain about the different shades of brown. One was so light it was almost white!

Ann thought a minute. "I guess a circus would have to have grown-up seals," she said. "You look down there and you'll see that the littlest seals are a real light brown. There's one lying in the sun that's almost white. He's a baby. He'll lose all that whitish hair and grow a new brown coat. Then, when he's even older, he'll lose the brown coat and grow a black one."

Betsy sighed. "I wish we could have that baby seal for a pet," she said earnestly. "Look at his big dark eyes!"

Ann laughed. "He'd hate it," she said. "He likes to flop around in the shallow water with his mother. She'll teach him things like hiding from the big storms and swimming against the tide and what's good to eat."

"B-r-r-r!" said Betsy. "I'm cold. Let's run." They picked up their coats and, after one last look at the seals, started to run toward the west side of the island. When they came to the salt swamp, Ann pointed out the narrow bridge of logs that crossed it, providing a shortcut. All around the edge of the swamp something bright red grew close to the ground. When they had crossed the logs, Betsy picked one of the bright small berries and bit it in two. "Oh my," she cried. "That's awful sour!"

"It's awful good," Ann answered, running on. "They're cranberries," she called over her shoulder.

Betsy knew what cranberries were. She had them in sauce on Thanksgiving Day, but they were fat juicy berries, not small and sour and dry. Even the cranberries were different on Bleak Island!

Along the shore they ran, past a pebbly strip of beach where the water was still and blue, to a point from which they could see a tangle of poles tied together and sticking out of the water. "What's that?" asked Betsy.

"We'll row out and look at it sometime," said Ann. "It's called a weir. It's a place for catching fish. It belongs to my father. It keeps getting broken by the storms, so it's a lot of work to fix it up again. See, there's my father now, out in his dory. He's tying the net back onto the poles, I guess. He can do anything," she said proudly.

The two little girls stood for a while on the grassy headland above the water, which rolled up in fat, slow waves. It was very still, for the wind was dropping and the sun was low, slipping behind the western hills on Big Island. Even the noisy seagulls were quiet. Some of them perched on top of the tall poles around the edge of the weir, staring at the man and his boat in a sleepy sort of way.

"When my father comes ashore," Ann said, "we can go up in the tower. I'm not supposed to go there unless he is there."

Betsy stepped back a little in a frightened way. "It's awfully far out in the water," she said in a small voice. "Would we go over that sort of bridge? What if we fell in?"

Ann turned and stared at Betsy. For a moment she didn't say a word. It was just as she knew it would be. Betsy was really truly a scaredy-cat. She was like the girls who came on picnics with the summer people—girls who were scared to jump in and out of boats and squealed when they stepped on seaweed or saw a jellyfish.

"You don't have to go in the tower, now or ever," she said at last. "You're just like a summer person's little girl!" Then she turned on her heel and ran as fast as she could to the tower, crossing the high little bridge on quick sure feet.

Betsy burst into tears. This place was horrible, she thought. Everything was new and strange. "I want to go home," she sobbed aloud. Then she went slowly up the path that went past Ann's house to Grandma Gates'. She rubbed her eyes on her sleeve, but the tears ran down her cheeks faster than ever.

She felt cold and homesick and awfully sorry for herself.

Just as she reached Ann's house, Mrs. Marsh came to the kitchen door. "Where's Ann?" she called.

"She's down at the tower," Betsy said. "I was too scared to go over that bridge." She paused and gulped. "I guess I'm awful scared of lots of things," she said miserably.

Mrs. Marsh ran lightly down the path to Betsy. She put her arm around her shoulder. "You come in and get warmed up by the stove," she said briskly. "It gets nippy the very minute the sun goes down." She glanced down at the little girl and saw the wet smudges made by tears on her cheeks, but she didn't mention them.

When they went into the kitchen, Ann's mother pulled a chair up to the stove and pushed Betsy gently down on it. Then she gave the little girl a thick china cup in which there was something hot and milky and sweet. Betsy sipped it slowly. It was very good.

"That's what my mother called 'cambric tea,'" Mrs. Marsh said cheerfully. "It's really nothing but hot water and milk and sugar, but it's good, and warming too." She looked down the window to the tower.

"The light's just gone on," she said. She watched it flash for a few minutes and then she said slowly, "Someday very soon you'll love the tower just as we all do. Right now it's part of a lot of strangeness you never saw before. You're not exactly scared. You're just surprised at a lot of things. Some of them you don't understand."

Then she laughed a little. "You should have seen Ann when I took her to Boston last year! She had never seen taxis and automobiles tearing through the streets. There were more people on Washington Street than she knew existed. She'd never ridden in the subway or an elevator or an escalator before. Why, she was shaking like a leaf at the end of her first day there. It was all new and strange to her, just as Bleak Island is to you. That's all."

Betsy finished her cambric tea. She felt lots better, but one thing still bothered her. "Ann doesn't like scaredy-cats," she said slowly. "Ann doesn't like me. She said I was like a summer person's little girl. What's a summer person?" she asked anxiously.

Mrs. Marsh stopped smiling. "Ann thinks the people who don't live here all the year round, but who come here for picnics, are queer and different. That's something I wish she'd change her mind about. People are people—they're grown-ups or old people or children, some of them friendly, some of them unfriendly. Their ways may be different if they live in different parts of the world, but they're the same inside. Perhaps I can get that into Ann's head in school this year."

"School!" said Betsy in a surprised tone of voice. "Where's the school? I didn't see a schoolhouse anywhere on this island."

Mrs. Marsh laughed. "It's right here in this kitchen," she said. "Most children of lighthouse keepers go ashore for the school year. But I taught school before I was married, so they let me keep school here for Ann until she's old enough for high school. Every year I get her books and the lesson outlines from the superintendent of schools on Big Island. And every June, Ann goes to Big Island to take some tests to see if she's ready to go into the next grade. I should think you'd be doing third-grade work, just as Ann is. You tell Grandma Gates to let you come down here tomorrow morning at nine, and we'll have school together."

Then she looked out the window. "Ann and her father are coming up the path now for supper. You'd better skip along to Grandma Gates'. If you hurry, you'll get there before it's very dark."

Betsy stood up and went to the door. "Thank you very much," she said shyly. "Thank you for the cambric tea and everything."

Just then Ann opened the door. The two little girls looked at each other silently for a moment. "It's getting dark," said Ann. "You'd better give Betsy a flashlight."

Betsy smiled. "I'm not scared of things I know about. I know about the path to Grandma's, so I don't need a flashlight." She paused for a minute, thinking. "Pretty soon," said Betsy, "I won't be scared of anything on this island."

School Is Fun!

THE VERY NEXT morning, Betsy ran down the path to Ann's house just before nine o'clock. She found that the big kitchen table had been pulled out into the middle of the room and that three chairs were placed around it. Ann and her mother were sitting by the table, on which books and paper and pencils were neatly laid out.

"This is your chair," said Mrs. Marsh, pointing to the chair at her left. "School will begin as soon as you have hung up your coat and scarf." Betsy hung up her things and slid into her chair quietly.

"We always begin with a little prayer," Ann's mother said. They bowed their heads and together said, "Our Father who art in heaven." The room was full of sunshine and warmth. It was the nicest schoolroom Betsy had ever seen.

The lessons that followed were the most interesting lessons the little girl from Ohio had ever had. The spelling words were words that belonged to the island, such as "lighthouse" and "buoy" and "lobsters." Though many of them were new words to Betsy, she was a naturally good speller and did very well. She began to feel quite proud.

Ann, however, did especially well in arithmetic, and Betsy gazed admiringly at her as she made neat little figures on her paper, adding and subtracting correctly. Both little girls loved to read and had a happy time reading to each other. Almost before Betsy realized it, it was noon.

"Recess!" cried Mrs. Marsh. "You stay for lunch with us, Betsy. Ann's father has gone to Big Island today, so there will just be the three of us. Tomorrow you can go home for lunch, coming back at one o'clock."

Ann and Betsy moved the books and papers and pencils away from the table and spread a bright checkered cloth on it. They put bowls and plates and spoons on the table. Then Mrs. Marsh filled the bowls with steaming-hot fish chowder. She put a small basket filled with polished apples in the middle of the table. From the oven she took a pan of hot biscuits. Then they all sat down to eat their lunch.

When lunch was over and the dishes were done, Mrs. Marsh helped them put on their coats.

"Now," she said, "go out into the sunshine until I call you!"

For about twenty minutes, the two little girls played tag in among the trees behind the house. Then they came in, rather out of breath, and settled down again at the table, which had become a school table once more.

Ann's eyes were shining. "This is the part of school I like best," she said.

"What happens?" asked Betsy.

"Mother calls it 'the world we live in,'" said Ann. "It's real interesting!"

"Do you mean just this world here on the island?" asked Betsy.

"We might start with the island," said Mrs. Marsh. "Ann knows a good deal about the island. Suppose you start telling all that you know about the island. You can think quietly about it for a few minutes, first."

Ann thought hard. Then she picked up a pencil. "This is the way the island looks," she said slowly. She drew a rather wobbly circle on the paper. Then she drew the lighthouse and her house and

Grandma Gates' house. She drew the rocks where they had seen the rocks. "The tide is out in my picture," she explained.

"What's the tide?" asked Betsy.

Ann explained how the sea slowly rises and falls all around the world, so that the waters creep up the shores for six hours and then fall away for six hours.

"Each day," Mrs. Marsh added, "the tide is one hour later in reaching its height than it was the day before and one hour later in reaching its lowest point. It is just after one o'clock now. The tide is what we call 'full' now. Tomorrow it will be full at two o'clock."

"Is there a clock in the ocean?" asked Betsy.

"No," said Mrs. Marsh. "The sea is following a law—what we call a law of nature. There are lots and lots of laws of nature. One controls the sun and moon and stars. Another controls the seasons. Others control the way things grow on the earth."

Ann was drawing busily while her mother talked. "I forgot to put the windows for the light in the lighthouse," she explained.

"It's awfully tall," Betsy said, looking at the picture.

"It's one hundred and fourteen feet tall," Ann's mother said. "This kitchen is about eleven feet wide, so the lighthouse is more than ten times as tall as this room's width."

"How do you get way to the top of the tower?" Betsy asked.

"You climb a lot of steps," Ann said. "But it's not too bad because there are sort of landings where you can stop and get your breath. Our island is too rocky at this end to dig a well, so the first stage you come to is an enormous big cistern."

"What's a cistern?" asked Betsy.

Ann began to look a little irritated by Betsy's questions, so her mother answered. "It's like a big room with no windows. It's full of hundreds of gallons of water. The Coast Guard tender comes out twice a year and fills it."

"The next stage is the engine room. There are two engines to run the foghorn. The minute the fog comes, my father starts the foghorn blowing, to tell boats to keep away from the ledges. It has a deep strong voice, and it sounds very brave on a dark foggy night." Ann paused. "It sounds friendly," she said.

Mrs. Marsh drew two lines across Ann's picture of the lighthouse. "This is the bunkroom," she said, pointing to the first line. "We used to have two assistants here, but now we are allowed to have only one. He's been on Big Island for two days now because his father is very sick. He'll be back tonight. The man on watch can rest in the bunkroom if it's a quiet night. Dick, the assistant, isn't married and he likes the bunkroom so much that he chooses to sleep there instead of at our house. He especially likes it on very stormy nights, when the big waves beat against the lighthouse and dash so high that their spray hits the windows at the very top."

"What's that?" asked Betsy, pointing to the second line Mrs. Marsh had drawn.

Ann leaned over Betsy's shoulder. "That's the watch room, where the radio and the telephone are. It's almost the most interesting part of the tower. On the landing above, there's the machinery that runs the lights, and just above that are the lights themselves. The little platform that runs around the lighthouse is just outside the room where the lights are. You can stand out there and see the whole world, every single bit of it!"

Betsy looked puzzled. "I've been up on the roof of the apartment house where I live in Ohio, and I thought I could see the whole world from there. I can see lots and lots of buildings—houses, churches, and stores. I can see the river. But I can't see the ocean at all."

"What's an apartment house?" asked Ann.

Betsy drew a big square on her paper. She drew lots and lots of windows. Then she looked at her picture for a long minute. "It begins underground," she said slowly. "It has a great big enormous cellar, with a furnace room in part of it. In another part there's a laundry, with a long row of washing machines and a place to dry things. And in our apartment house, because a lot of children live there, there's a playroom and a place to ride your bicycle on rainy days. The elevator starts down there; it's the kind of elevator you can run yourself if you are grown-up."

"I'd like to do that," said Ann.

"There's a big hall on the first floor," Betsy went on. "In one wall there are lots and lots of mailboxes. Each family has a key to its own box. And in one corner there's a place where the telephone operator sits. When anyone wants to telephone us, he calls the number of the apartment house, and the operator connects the call to the telephone in our apartment."

"If there are lots and lots of people there," said Ann, "is it like a lot of little houses piled up on top of one another?"

Betsy laughed. "Not real little houses," she said. "But each family has four or five rooms on the same floor. They're kind of connected, but they're private. They have an outside door into the hall where the elevator is and another into the back hall where the stairs are. When you go into your apartment and shut the door, it's really truly yours. The rest of the building belongs to everybody who lives there—even the flat roof where I go with my mother after supper in hot weather."

Mrs. Marsh began picking up the books and papers. "I think we learned something important today," she said. "We learned that this is an awfully big world. Ann thought she could see all of it from the tower, and she thought the sea was something everyone could look at. Betsy thought she could see the whole world from the roof of her apartment house, and I guess she thought most of the world was full of buildings, the way a city is. Each of you has seen a little bit of an awfully big world, and there's lots more to see, all of it interesting. Now," she said, going to the pantry, "school's over. It's time for cookies and milk. Then you can go out to play."

Betsy drank her milk slowly. "This has been the best fun I've ever had in school in my whole life," she said happily.

"I'd like to run that elevator all by myself," said Ann.

Then they put on their coats and ran out to play.

7
Betsy Tries Hard

AT THE END of Betsy's first week on Bleak Island, she had begun to feel more at home. She and her grandma had become very good friends, and she loved the old lady with all her heart. Mrs. Gates had a storehouse of wonderful stories about the sea tucked away in her memory. Almost every evening, when Betsy was all ready for bed, she sat in the big chair that had been her grandpa's, with her feet toasting gently on the low rim of the stove, while Grandma told her a story. It had not been hard to learn the things Grandma asked her to do in a special way—such as making her bed and polishing the kitchen windows with a soft cloth when they steamed up on a cold day.

Betsy had also grown very fond of Captain Marsh, who teased her just as he did Ann. From him she learned a great deal about his work, which, she found, was much more than simply turning on the lights and the foghorn. She knew the careful plans that had to be made so they would never run short of food or supplies. She discovered that every bit of Ann's house, as well as the tower and the oil house, must be kept gleaming and freshly painted, for they belonged to the United States government and were regularly inspected. It was not long before she decided that Captain Marsh was a wonderfully good and wise man, as well as being lots of fun.

Mrs. Marsh had been her friend from the very first, and, as she was young and full of fun, she kept Betsy from missing her mother too much.

Ann was—well, Betsy still couldn't figure Ann out at all. In school she was just the kind of friend Betsy had always wanted. In lots of ways, the hours of school were the happiest time of the day for Betsy. But on Saturdays and Sunday afternoons, or during the play hours after school, Ann still had moments when she acted queer and almost cross. When Betsy went to bed at night, she thought a lot about Ann, trying to figure out why she was so different at different times.

"It's because I'm a scaredy-cat," she thought unhappily. And she would resolve to be just as brave as Ann was, the very next day.

The next afternoon, Betsy came home to Grandma Gates saying that her throat hurt. Grandma told her to say "a-a-h" and looked down her throat.

"H-m," Grandma Gates said thoughtfully. "It looks real red and sore. No wonder it hurts. I'll just fix you some ginger tea while you get undressed and pop into bed. Then I'll put a rag around your throat and soak the misery out. You put on your flannel nightgown, and wear your bathrobe right into bed, so you'll be real warm. We'll fix you up in no time."

By the time Betsy had crawled into bed, she felt very achy and miserable. She tried hard to swallow the hot ginger tea, which Grandma fed her with a spoon, but it hurt going down. The "rag" Grandma had spoken of turned out to be a strip of soft old flannel soaked in kerosene. When it was wrapped around her throat, it felt warm and good, but the smell of kerosene was so strong that it made her eyes water. The next thing Betsy knew, she was crying, and crying hard. "Oh dear, oh dear," sobbed Betsy, "I want my mother!"

Grandma Gates sat down by the bed and began to knit busily. "I'm making you some nice red wool stockings to keep you warm when winter comes," she said cheerfully.

"I want to go back to Ohio," cried Betsy. "I don't want to be here all winter long. I want to go home!"

Grandma looked at Betsy over her glasses. Then she went on knitting, as if she were thinking of nothing in the world but the red stockings she was making. Betsy pulled the sheet and blankets high until there was nothing to be seen of her but one little brown pigtail sticking out from under the covers.

All of a sudden, something landed, kerplunk, on Betsy's feet, almost as if it had dropped from the skies. She threw back the covers with a squeal of surprise and stared at the foot of her bed. There stood the big mother cat, looking at her. Betsy moved her feet away from the middle of the bed and the big cat turned around three times and curled up in a big ball. Then Betsy heard a small mewing sound.

Grandma smiled. She leaned down and picked up the kitten and put it on the bed. It curled up beside its mother. "Lady Topside," said Grandma, "you ought to be ashamed of yourself. You know better than to jump on beds."

Betsy put out her hand slowly and stroked the kitten's soft fur. It began to purr like a small furry teakettle. It was a nice sound.

"Animals are a lot of company on an island,"

Grandma said. She was quiet for a moment, as if remembering something. Then she went on, in a storytelling kind of voice. "They can get in a heap of trouble, though," she said.

Betsy's eyes felt hot and heavy, so she closed them. But she wanted to know what trouble animals could get into, so she said hoarsely, "Tell me about some of the trouble."

"Well," said Grandma, "there was the cow that fell down the cliff."

Betsy's eyes popped wide open. Grandma smiled. "That was the stubbornest cow that ever was born. It belonged to a lightkeeper off to the south of us. He tried tying her up and fencing her in, but that cow was just set on going her own way, free and easy. And she ended up by sliding right down a cliff twenty-five feet high. That's almost three times the height of this room. She wasn't hurt much, but she let out a mighty bellow, mostly because she was so mad. The keeper and his helpers came running and looked down at her, far below and near the high-water mark. She had scrambled to her feet and was standing there, mooing her head off. The keeper shouted at her, and she turned around and glared at him. It was just as if she were asking him what he was going to do about the fix she was in. He had to laugh—she looked so cross and foolish, as if it were all his fault. But he knew that in three hours the water would rise almost up to where she was standing, and if he didn't get her off, she might do something even more crazy, like trying to swim to the mainland seven miles away."

"Wasn't there any path to the shore?" asked Betsy.

"Not a smidgin of a path," said Grandma firmly. "The men stood looking and figuring what to do for a spell. Then the keeper sent the helpers back to the lighthouse for boards and a lot of rope. When the gear was brought, they let it down slowly to the rocky shore below. Then the youngest helper, a boy named Ben, went down the rope, hand over hand. The keeper shouted directions down to Ben, who carefully lashed four or five boards together. Then he fastened a rope to each corner. 'Now get that stubborn cow aboard,' shouted the keeper.

"Wouldn't you think," asked Grandma, "that any cow would want to get off that rocky ledge with the water creeping nearer every minute?"

Betsy nodded.

"Well," said Grandma, "Ben pushed. He coaxed. He shoved. He talked to that cow as if she were human. At last he had to put a rope around her neck and drag her onto the planks. He wanted to get off, thinking the rope wouldn't be strong enough to hold him and the cow, but the minute he stepped off, the cow stepped, too. So at last the keeper and the other two men had to haul Ben and the cow up together, the cow mooing every foot of the way and stomping her hoofs on the planking as if she didn't want to be rescued at all. When she was safe and sound on the headland, the keeper locked her up in the woodshed for the night, and the very next day he sent her ashore on the Coast Guard tender and swapped her for a cow that would stay put."

"Oh my," said Betsy sleepily, "tell me some more." She was beginning to feel very warm and cozy. Grandma's voice seemed to come to her from far, far away.

"Then there was the dog they used to tell about that was as smart as a man," Grandma went on. "I can't vouch for how true it is, for I never laid eyes on the dog or the keeper who told the story. But it was common talk all up and down the coast thirty years ago."

"Thirty years ago . . . thirty years ago," echoed in Betsy's ears sleepily. Long, long ago.

"It seems there was a real heavy fog one day. The lightkeeper was on the mainland getting supplies, and he got caught in the fog coming back and couldn't get his bearings. The keeper's wife went down to the shore to listen for the sound of her husband's boat engine; she couldn't hear anything but the angry roar of slow, big waves piling up on the reef. After a while she went back to the house, tired and cold from the snow that had begun to fall through the cloud of fog. The first thing she saw when she came into the house was Rover, the golden retriever, who was lying in front of the stove, fast asleep. Now the keeper had taught Rover to jump up and pull the rope that made the old-fashioned fogbell ring. Rover was so proud of this trick that he performed it whenever he heard the engine of the keeper's boat putt-putting along the shore. He had such keen hearing that he could tell the sound of his master's boat from that of any other.

"The tired woman looked at Rover. Then she opened the door and pointed out into the darkness. 'Rover,' she said. 'Go out and listen for the boat. Then ring the bell.' Then she sat down by the stove to get warm.

"Rover ran down to the shore. He sat down on a big rock and listened to all the sounds in the darkness. He could hear the sea thundering on the reefs, but there was no putt-putt of an engine to be heard. The snow covered his curly hair with a thick coat, and the wind blew spray in his face, but he sat there listening for a long time. Finally, he went back to the house and scratched at the door. The keeper's wife let him in and brushed off the snow, rubbing him down with a rough towel. Then Rover lay down by the fire until he was warm again.

"Suddenly he jumped up and ran to the door, whining. The woman let him out and stood at the open door for a minute, listening. She could hear nothing but the whine of the wind and the booming of the sea on the ledges. Rover had heard something, though! He ran as fast as he could to the fogbell and jumped up to seize the rope in his teeth. But the snow was wet and slippery. Over and over again he jumped, but each time he slipped back, and the rope still dangled free. After a dozen or more tries, Rover ran down to the shore and stood there a moment, listening. Yes, there it was, faint but unmistakable, the putt-putt-putt of his master's engine. It came a little closer every minute. But it was moving to the south. If it kept on, it would pile up on the reef and be smashed to smithereens.

"Off he dashed to the rocky headland just above the boat landing into which his master could come safely. Then he braced himself, with his legs spread wide against the blowing snow and fog, and began to bark furiously.

"The lighthouse keeper had been nosing his boat through the waves carefully. He could see the flashing light, with big snowflakes dancing in front of it, but he knew there was only one spot where he could land safely in a storm. He listened for the fogbell to guide him. Suddenly, out of the darkness,

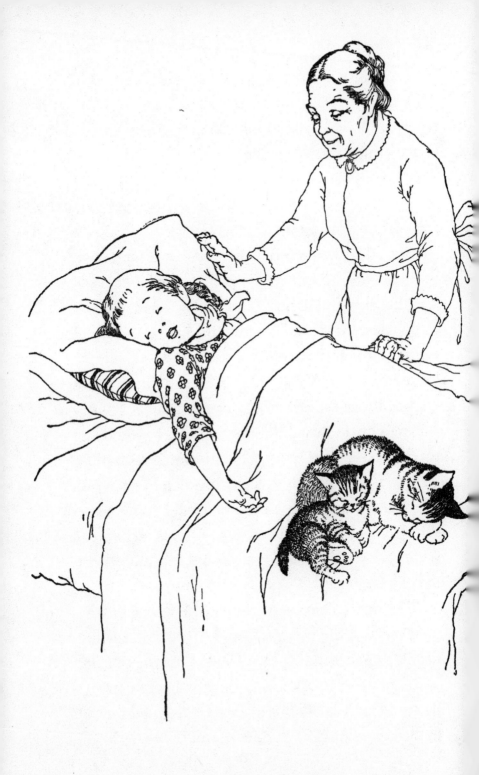

came Rover's bark. It seemed to say, 'Come this way! Come this way!'

"'Ahoy there, Rover!' he shouted, and swung the boat around, heading for the sound of the barking. No music, no other sound, had ever sounded more beautiful to the cold and tired man than that hoarse and excited barking.

"In a few minutes, he reached the landing. Rover jumped into the boat and covered the keeper's hands with wet kisses. As soon as he had pulled the boat up out of the reach of the tide, the keeper and Rover hurried up the path to the warm, brightly lit house."

Grandma stood up and put her knitting into the pocket of her big apron. She went over to the bed and looked down at Betsy. The little girl and the big cat and the kitten were all fast asleep. She put her hand on Betsy's forehead. It was no longer hot and dry, but cool and damp.

"That's good," said Grandma. "The fever's broken. You'll be fit as a fiddle tomorrow morning." She leaned over and kissed the little girl gently on the cheek. Then she scooped up the big cat and put out the light. "Come on, Lady Topside," she said. "It's time us two old ladies had some supper and went to bed." Then she went quietly downstairs.

8
What's a Line Storm?

THE CRISP BLUE days of September slipped by rapidly. Perhaps, Betsy thought, the cold air, only faintly warmed by the sun at noon, made everyone feel excited, as it did her. Certainly, all hands were busily at work. Captain Marsh and Dick painted every inch of the keeper's house inside and out so that it was shining clean. They painted the lifeboat and the dory, which they had pulled high up on the shore. Extra supplies and oil for the lamps were carried up into the tower and stowed away. Grandma Gates cut back the roses that had climbed over her picket fence, and, squatting in the sharp-edged grass, painted each picket a dazzling white.

Mrs. Marsh and Ann took winter clothes down

from the attic where they had been stored and hung them out on the line so that the smell of moth flakes would blow out of them. They cleaned closets and pantry shelves. One day, instead of the usual school lessons, Mrs. Marsh made what she called an "inventory," counting up all her food supplies and her stacks of dish towels, sheets, pillowcases, and other necessities. Ann and Betsy made neat lists for her. Ann's list began: "We Have." Betsy's began: "We Need."

As they wrote, Ann looked up at her mother. "It's almost time for the Wishing Book to come, isn't it?"

"The Wishing Book!" said Betsy. "What's that?"

Mrs. Marsh laughed. "That's what Ann calls the big mail order catalog that comes every fall. After I've made out our order for things we really need, Ann has a great time making out lists for things she wishes for. It's a nice thing to do on snowy days when it's wet and cold outside. We use it in school, too. You wait and see!"

Ann smiled in a grown-up way. "When I was seven," she said, "I wanted to do the things boys did. I wanted to go hunting and fishing on one of the big lakes up north on the mainland. I made great big long lists then, I can tell you—tents and hip boots and fishing rods and guns and all sorts of food. I don't want to be like a boy anymore, though."

"I know what I'd do if I were a boy," Betsy said. "I'd go to the Coast Guard school and be a lightkeeper." She looked down to the shore where the man from the Coast Guard was unloading coal and carrying it up to the storehouse. Captain Marsh was standing on the little catwalk that ran around the tower, looking down, laughing and joking with them.

"Ho," laughed Ann. "You couldn't be a lightkeeper. Lightkeepers are brave."

For a moment Betsy's lip quivered. *Would I be scared?* she thought. *Why am I a scaredy-cat? I don't want to be.*

Mrs. Marsh shook her head at Ann. "That's not very kind, Ann," she said reprovingly. "There are lots of ways of being brave besides being a lightkeeper. You'll find out someday."

Ann said nothing but went back to making her list.

"Mrs. Marsh," asked Betsy, "why is everyone working so hard and fast, as if they were in a hurry? Even the Coast Guard men are hustling to get the coal off instead of visiting, as they often do."

"It's a busy time of the year," Mrs. Marsh said. "Pretty soon it will be winter, and everything must be ready for the days when not even the Coast Guard boat can land here. We'll be snug and tight when those days come, but it takes a lot of work to get ready for them."

"And before that," said Ann, "there's the line storm."

Betsy was just going to ask, "What's a line storm?" when she caught Mrs. Marsh's eye.

Ann's mother spoke quickly. "A line storm comes twice a year, in the fall and in the spring. The spring storms aren't much around here, but the fall storms are hard on everyone, especially the lobstermen and fishermen. Traps are lost if they are left out in the storm, and boats are torn from their moorings. Line storms mean very high tides, strong winds, and big waves. They last two or three days, and we're all glad when they're gone."

"Remember last March?" asked Ann, looking up.

"You mean when Sam Lurvey's boat went ashore," said Mrs. Marsh, nodding her head.

"Did it go ashore here?" asked Betsy.

Mrs. Marsh went to the window. "Look out here, Betsy," she said. "See that long rocky point on the island to the southwest of us? That's Goose Head. That's where the boat went ashore. Sam Lurvey knows how to manage in any kind of boat except a boat with a stalled engine, and that was what he had, just as he started around Goose Head. It was about four o'clock in the afternoon when the engine died, and all the other lobstermen had come in. There wasn't a boat in sight.

"Of course," she went on, "he had a ship-to-shore radio so he got the news of his difficulty to the Coast Guard depot right away. He knew they'd be out soon, so he just settled himself to wait. His boat had run fair and square across the ledge, so it didn't rub any, or move, but the big waves slapped against it hard, and he had to keep his pump going to get the water out of her.

"It wasn't long before the picket boat (that's a real quick-moving, able boat about forty feet long) came out, and the men on her tried, with ropes and boathooks, to budge the *Nancy*, Sam's boat. She wouldn't move an inch. After trying for an hour or more, they asked for help, and the depot sent out a tugboat. The tugboat crew had brought timbers with them. They wedged the timbers under the *Nancy* and inched her gently off into the water. Then they towed the boat in through the darkness."

"Was it all smashed on the ledge?" asked Betsy.

"No," said Mrs. Marsh. "That was the strange thing about the whole affair. The paint was scratched some along the hull, but otherwise it wasn't hurt at all where it had rested on the ledge. But the pilothouse, the place where Sam stood to steer, was smashed in just like an eggshell. Those big waves pounding down on it crushed it all to smithereens."

"Goodness," said Betsy, chewing her pencil. Then she asked thoughtfully, "Will the line storm come soon?"

"Any day now," said Mrs. Marsh. "You can tell when it is coming. The sky will be hazy and smoky. We may have a real warm day or two, like July weather. The ocean will have a slick, oily look, and everything will be still—no wind, no surf. Even the seagulls will quiet down."

"And then," said Ann, "bim, bam, bam! Everything happens at once—great big waves, surf a mile high, and a wind that just screeches! It's exciting."

Betsy frowned a little. It sounded more scary than exciting. She wasn't at all sure she liked the idea of a line storm. But she wouldn't show she was scared. She didn't say a word.

The very next day was warm and still and smoky. The little waves that before had merely ruffled the sea slid into one smooth gray-looking roll. Betsy helped Grandma Gates tie the lilac bush back against the house so that, when the wind came, it wouldn't beat against the parlor window and break it. There was no school that day, for Ann and her mother were busy doing last-minute chores before the storm came.

Once, Betsy asked her grandmother what they would do when the storm came. "We'll sit indoors," said Grandma Gates, "snug as a bug in a rug. We'll read and sew and listen to the radio and cook. You'll have to take care of Sally. Her mother, Lady Topside, is old and wise about storms. But Sally is only a kitten. She may want to go out and play in it. We'll be just fine."

Betsy stared at her grandmother. She didn't sound a bit worried or frightened. Maybe there wasn't anything to be scared of after all. "Aren't you scared at all?" she asked in a small voice.

Grandma put her arm around the little girl. "I'm an old woman," she said slowly. "I've seen an awful lot of storms—rain, hail, lightning, and bad winds. I've known the time when men didn't have radios and telephones on lonesome islands. I've seen boats smashed up like matchboxes, and I've seen men who were drowned. But I've seen other things too. The sunrise over the ocean is one of them, and the path the moon makes across the water—silver and black—is another. I've seen the way a young boy looked when he was rescued and brought home safe to his mother. Everything balances up in this world, Betsy. Good and bad—though I've seen more good in the long run. You just never can tell where the good will be. It crops out when you don't expect it or look for it. No," she said quietly, "I'm not scared. I'm not scared one bit."

When Betsy woke up, two days later, and heard the sea roaring and the rain pelting against her window, she lay very still and thought of what Grandma Gates had said. "It's here," she said in a whisper. "The line storm is here. I'm not scared. I'm not scared one bit." Then she got dressed and went downstairs for breakfast.

9
Come Kitty, Come Kitty!

FOR TWO DAYS Betsy and Grandma stayed indoors, with the windows shuttered and the doors barred. The rain beat on the roof and seeped in a little under the kitchen door. The wind tore at the house as if it wanted to blow it apart. Every now and then, a shingle would be ripped off the roof with a tearing sound. The trees in back of the house groaned and creaked as if they hurt. By suppertime of the second day, Betsy was sick and tired of the noise the storm made. The little girl felt suddenly very far away from her father and mother. "I wish I were back home," she said in a choking sort of voice.

Grandma Gates looked at her quickly. "I know," she said understandingly. "This wind blows something tedious. The only thing for us to do is to change things around somehow so that there'll be a difference."

Betsy stared at her grandmother. What did she mean, "change things around"? Was she going to move the kitchen furniture the way Betsy's mother moved furniture sometimes? The stove and sink and shelves were fixed to the floor and walls. There was nothing to move but Grandma's rocking chair, the table, and the chairs.

Grandma laughed when Betsy told her this. "No, child," she explained. "We won't change the kitchen around. We'll just change you around. You get your nightclothes, and bring them down here. Then you can sleep with me in my big double bed. Two's company on a night like this. I'll go to bed when you do, and we'll be real snug."

When Betsy woke up the next morning, she felt almost happy. She had been warm and safe all night long, curled up beside Grandma. Besides, two days of the storm were gone. It couldn't last much longer.

Although the surf still made an angry noise, and the wind blew very hard, by the time Betsy and her grandma had finished doing the dishes and making the bed, a faint light shone behind the low, heavy clouds. It must be the sun! Betsy ran to the kitchen door and drew back the bolt.

"Betsy," called her grandma from the front room, "why don't you work on the quilt for your doll? I'm tidying up the things in this old chest, and I may find some pieces for you."

Sally jumped down from the rocker and ran to Betsy, rubbing against her ankles. Betsy picked up the kitten. "I'm playing with the kitty," Betsy said.

Then, holding the kitten with her left hand, she slid back the bolt and very carefully opened the door. She stepped out on the broad rock step and shut the door quietly behind her. The wind blew in her face, but the door at her back broke its force.

"Oh my," said Betsy. "It's nice out-of-doors!"

Just then the kitten jumped from her arms and went skittering down the path toward Ann's house. There was nothing for Betsy to do but to run after it as fast as she could. But the kitten ran faster. When

Betsy reached the keeper's house, she banged on the kitchen door.

"Come help me," she shouted. "Kitty got out!" Then she ran on. Now that she was away from the house, the wind tore at her hair and clothes. She had to push hard against it to run at all. But run she did, calling as loud as she could, "Sally, Sally," and, "Come, kitty! Come, kitty!" But the wind seemed to take the words right out of her mouth.

Sally paid no attention. She just streaked along, a small black furry thing, so close to the ground that the wind could not stop her. Down the path she flew, making straight for the little bridge that led to the lighthouse tower. The bridge, which had no supports under it but stretched from the side of the tower to supports on the shore, swung from side to side, squeaking crossly in the wind. When Betsy came to it, she stopped short and looked down. Far below, the water swirled and splashed and rushed, a black stream of icy cold water. Spray filled the air as big waves broke against the tower.

"I can't, I can't," sobbed Betsy. Then she found herself gripping the handrails on each side of the bridge and pulling herself across it against the wind. The bridge moved under her feet, and the little girl felt suddenly seasick, just as she had done the day she came to Bleak Island on the Coast

Guard boat. "I'll never get there!" Betsy cried to the wind.

But she did get there. Once inside the tower, away from the wind, it wasn't so bad, though whenever a big wave struck against the outside wall, the whole building seemed to shiver.

"Sally!" called Betsy at the foot of the steps.

"Sally!" echoed the walls back to her.

There was nothing to do but climb the steps—from one stage to the next—until she reached the watch room. She half expected to find Dick there, but he was on the stage above, cleaning spray from the windows around the light. But Sally was there, running around and around, crying and mewing.

"Don't cry," said Betsy. "Don't be afraid. There's nothing to be afraid of." Then she sat down on the floor and called gently, "Come, kitty." Slowly, slowly, the little black kitten edged around the room to Betsy, who sat very still, talking to the kitten in a soft voice. "You're all right, Sally," she said. "I'll take care of you. It's only a storm. Storms come, and they go away. It won't hurt you."

The kitten crept into Betsy's lap and lay still, shivering a little now and then as the little girl stroked her wet fur. When she began to purr, Betsy got up very carefully. She took an old woolen pea jacket from a peg on the wall and wrapped it snugly around the kitten. Then she started down the steps.

When Betsy reached the bridge, she looked across. There on the shore stood Captain Marsh and his wife, and holding fast to Ann's hand was Grandma, her hair covered by a dark shawl. "Want me to come get you?" shouted Captain Marsh.

Betsy shook her head. "We're all right," she shouted back. "We're not scared a bit." Then, holding the kitten firmly in her left arm, she took hold of the guardrail with her right hand. She didn't look down at the water at all, but bent her head to whisper comfortingly to the kitten. Then she walked slowly and carefully across the bridge to the shore.

10

Let's Be Friends

TWO DAYS LATER Bleak Island lay green and peaceful under a sunny sky, with the sea making a small white ruffle all along its shore. If it hadn't been for the driftwood and the tangles of seaweed, thrown high on the shore by the high tide, and the broken shingles lying on the ground, you'd never know a storm had come and gone. School had begun again, and after lessons were done, Betsy and Ann made popcorn balls to celebrate the end of the storm.

The popcorn balls were delicious, the kitchen was flooded with afternoon sunlight, and Betsy suddenly realized she was completely and absolutely happy. She licked molasses from her fingers and looked around her thoughtfully. "This is a nice place," she said slowly. "It's different, but it's nice." Then she took another popcorn ball to eat.

Yes, it was true. She was happy. There had been a bad storm, but it had gone away. She had been scolded by Grandma for going out-of-doors and letting kitty out, but Grandma loved her, she knew. Besides, she knew perfectly well when she opened the kitchen door that she was doing wrong. She had been awfully scared, but she had found the kitty and would never be scared of anything again. And today, in school, she had asked lots and lots of questions, but Ann had not acted scornful or superior at all.

As a matter of fact, Ann had very little to say. She had done her lessons quietly, without showing off at all. She had been thinking hard about something that had nothing to do with lessons. Once in a while, she would look up from her work and stare unblinkingly at Betsy, as if she were a complete stranger. When the little girl caught Ann looking at her in that studying kind of way, she just smiled quietly. There was nothing to be afraid of or troubled about.

"Yes," Betsy said again. "It's nice here, and it is different."

Mrs. Marsh spread out a big sheet of wax paper. "Betsy," she said, "why don't you and Ann cut out some squares to wrap the balls in? Then you can take some down to Captain Marsh and Dick. When you go home, you may take some to Grandma."

As the little girls cut and wrapped, Mrs. Marsh sat down near them. She began to crochet. "You just said a very grown-up thing, Betsy," she remarked thoughtfully. "It would be a pretty dull world if there weren't any differences. There need to be places like Ohio and places like Bleak Island and New York City and little towns way out on the prairies to make up a country. And there have to be lots of countries, different in lots of ways, to make up a world. The only trouble comes when people say, 'I don't like it because it's different,' or, 'I'm afraid of it because it's different.'"

Ann looked up from the popcorn ball she was wrapping carefully. "That's what I said about Betsy, at first," she said slowly. "I said she was summer people, and I didn't like summer people." Then she looked at Betsy and smiled. "You're not like summer people at all," she said.

Mrs. Marsh drew a long breath. For a minute she tried to think what to say that would get Ann to stop thinking and talking as she did about summer people. Before she had spoken, Betsy asked Ann a question. "What are summer people like, anyway?" she wanted to know.

Ann frowned. She thought hard. Then she said in an embarrassed tone of voice, "I don't know."

Mrs. Marsh burst out laughing. She gave Ann a little pat on the shoulder. "That's my honest little girl," she said. "You used to think you knew all about them. But next summer we'll walk out to the Point when they come, and you'll see they're just people. Some may speak a little differently from us, but there's nothing wrong in that. Some may make mistakes, like offering your father money for showing them the tower. But they are only trying to say thank you that way. A lot of them will ask what seem like foolish questions—"

She was interrupted by Betsy, who laughed a little, shyly, but happily. "Like me," said Betsy. "Just like me. But if you've never been on a lighthouse island, you want to know all about it. And how can you find out unless you ask questions?"

Ann put the last popcorn ball on the top of the pile. Then she reached out her sticky hand and put it on top of Betsy's. "Let's be best friends," she said softly. "Let's always be best friends." Then she picked up the big platter of popcorn balls. "Let's go down to the tower."

Side by side, the two little girls went down the path. When they came to the bridge, Betsy went ahead of Ann, without pausing for even a minute of fear. She ran across, and when she reached the tower, she turned and called back to Ann, "Come on, scaredy-cat!"

Ann laughed so hard that the topmost popcorn ball rolled right off the top of the pile and fell, plop, into the water. In a second a seagull dived down to the white ball floating on the water, screaming with excitement over this new kind of fish.

For a minute Ann and Betsy stood in the tower door and looked at the great stretch of blue water and sky, with the mountains of Big Island in the distance.

"It's awfully nice," said Betsy again. "It's different, and it's awfully nice."

Ann smiled. "Someday maybe I can come to see you in Ohio," she said. "That's different, isn't it? Different and awfully nice."

Then the little girls went into the tower.

ELSPETH BRAGDON had a grandmother who hailed from Damariscotta, Maine. She herself spends as much of the summer as she can on a small island off the Maine coast, where she watches with delight the pleasure her own very young granddaughter takes in the sights and sounds and smells of Maine.

Mrs. Bragdon is the author of two books of light verse and a book of short stories for children.

MARJORIE TORREY had been drawing pictures as far back as she can remember. She studied at the National Academy of Design in New York and the Art Students League. Now she lives in California in a rambling house she and her husband built on a hilltop overlooking the San Fernando Valley. She is the author and illustrator of *The Merriweathers*.